ROBOT-PROOF YOURSELF

How to Survive the Robocalypse and Benefit from Robots and Automation

JASON SCHENKER

FI THE FUTURIST INSTITUTE

PRESTIGE PROFESSIONAL PUBLISHING

CONTENTS

INTRODUCTION

ROBOCALYPSE NOW?

Like Hansel in the movie *Zoolander*, robots and automation are "so hot right now." Books, articles, and television segments on automation, robotics, universal basic income, and the future of work are coming to dominate the *Zeitgeist* — the spirit of our time. People are becoming increasingly aware that automation and robots are going to be important — and unavoidable — parts of our working and non-working lives.

And, like you, many people wonder: Will robots take all the jobs? Will there by a Robocalypse?

When people talk about robots and the future of work, they tend to couch the outlook in one of two ways. At one end of the spectrum is the vision of Robocalypse: an apocalyptic future explicitly caused by robots, automation, and artificial intelligence. At the other end of the spectrum, is the vision of Robotopia: a heaven-on-earth future world of leisure, where all work is done for us by the machines.

Robocalypse Reality: Low-Education Jobs Will Disappear

Many jobs will disappear in the coming decades. And those jobs that require only low levels of education are at a very high risk of Robocalypse. A report issued by the Office of the President in December 2016 titled, "Artificial Intelligence, Automation, and the Economy," included data about low-income and low-skill jobs seen in the figure below.[1]

According to the report, 44 percent of jobs that require less than a high school diploma are "highly automatable," while zero jobs that require graduate degrees are "highly automatable." Zero. Plus, only one percent of jobs that require a Bachelor's degree are viewed as highly automatable. This graph shows the value of education in protecting yourself from being a victim of Robocalypse.

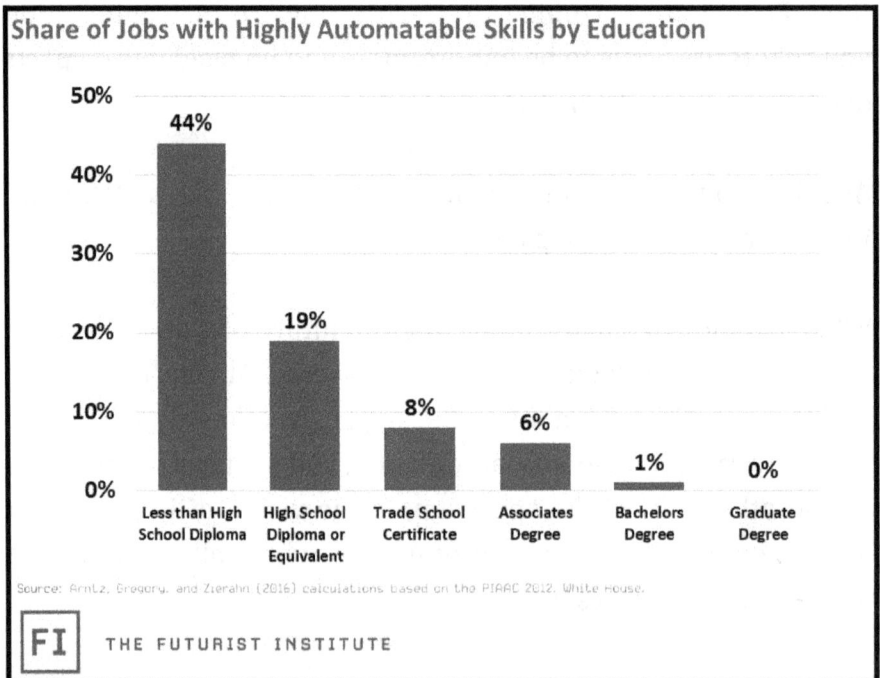

Share of Jobs with Highly Automatable Skills by Education

Less than High School Diploma	High School Diploma or Equivalent	Trade School Certificate	Associates Degree	Bachelors Degree	Graduate Degree
44%	19%	8%	6%	1%	0%

Source: Arntz, Gregory, and Zierahn (2016) calculations based on the PIAAC 2012. White House.

FI THE FUTURIST INSTITUTE

Be Prepared to Robot-Proof Your Career

In this book, I present recommendations for individuals to help them reap the rewards as we transition from the information age to the automation age. Workers need to seek out evergreen professional opportunities wherever they are, as well as embrace education and skill-building opportunities.

Getting more education — no matter what professional level you are at — will no longer optional. The jobs of the future will require continual learning, and the workers of the future that want to reap the benefits of the automation age will keep learning.

Without more education and skills — and a willingness and preparedness to adapt at a more rapid pace — you will be left behind. Trucking, transport, and retail could become professional wastelands. But healthcare, automation, information technology, and project management are likely to be professional promised lands.

In-Hand Classroom

The greatest defense against a future Robocalypse is an educated and adaptive workforce. We need to bridge education and skills gaps, and the in-hand classroom is a critical tool to help people survive and thrive during the coming robotic and automation changes to the labor market. Fortunately, access to online courses, certifications, professional designations, Bachelor's degrees, and Master's degrees has rapidly expanded. And the future of online education is likely to expand and accelerate further.

The Futurist Institute

One of the most important developments coming out of the process of writing *Jobs for Robots*, attending RoboBusiness, and learning about robots and the future of work, was the decision to found The Futurist Institute. After a long discussion with Martin Buehler, the head of Disney Imagineering R&D, about the challenges in developing talent in robotics, it occurred to me that most economists, analysts, and other business professionals are woefully unprepared for the changes that are coming to their fields.

The Futurist Institute is designed to help analysts and economists become futurists and long-term analysts. So far, I have created five different courses, designed to help analysts and economists become futurists: Futurist Fundamentals, The Future of Work, The Future of Data, The Future of Finance, and The Future of Transportation.

The Robot and Automation Almanac—2018

In addition to establishing The Futurist Institute, and a series of online training courses, my team and I are compiling articles from robot and automation experts to create an annual outlook for the year ahead called *The Robot and Automation Almanac*. The inaugural edition will be released in January 2018.

Need for More Training to Prepare People

The need for professional education outlets is rising — but there are still significant needs that have yet to be met. In taking a page from my own book, I have become significantly more involved in professional and online education since the release of *Jobs for Robots*. Since the publication of the first edition in February 2017, I have released a course with LinkedIn Learning on Financial Risk Management — and I have a number of other courses in development with LinkedIn Learning about critical U.S. and global economic indicators.

I was also inspired by my experience at RoboBusiness 2016 to personally dedicate a significant amount of time and resources to found The Futurist Institute™, in order to help analysts and economists become futurists. Through our program, which was developed over a 12-month period, an individual can become a Certified Futurist™ by taking courses in one of four professional tracks: standard, legal, accounting, or financial planning. We have even established a formal certification, so people can gain the skills and recognition to use the Futurist and Long-Term Analyst™ — or FLTA™— designation.

The need for professional skills development is high, and you should expect to see a lot more online learning offerings from professionals, like myself, who want to help others develop skills to move up the intellectual capital curve to become robot-proof.

ROBOT-PROOF
YOUR CAREER

You need to prepare for the acceleration of automation and robotics. You need to be robot-proof. But to be on the right side of these changes, you need a strategy to lay the groundwork of your own Robotopia.

There are three main strategies you can implement to prepare for the inevitable disruption of the labor force by automation:

Work in an Evergreen Industry: *Gain professional exposure to a career that will be in demand in the automation age.*

Learn Valuable Skills: *Take advantage of formal and informal education opportunities. Be prepared to learn more.*

Keep Moving: *Put yourself in a position to find opportunities, by changing industries, companies, or geographies.*

Let's look at examples for how these strategies can help you find upside, even if there are downturns in certain industries.

Strategy #1: Work in an Evergreen Industry

No matter how big the economic or automation risks, there are always some industries that will be evergreen. Make sure you have professional exposure to one of them.

The prospect of Robocalypse presents significant uncertainty for a number of industries, especially manufacturing and transportation. Meanwhile, other industries have a relatively low potential for automation. These include jobs in information technology, healthcare, and management. I discussed these dynamics in the book *Jobs for Robots*, and a full list of different industry exposures to automation.

Information Technology

Obviously, increased automation, robotics, and reliance on technology present significant opportunities for careers in the information technology space.

I recently had a call with a former colleague and good friend who works for a company at the forefront of vehicle automation. I was talking to him about the risks to the economy and financial markets. He was very concerned, because part of his compensation is tied to the performance of the NASDAQ.

I told him, "You have nothing to worry about." Then, he asked me why. I was laughing as I told him, "The stock market may fall, but you're going to be the last guy on earth with a job. Literally. The. Last. Guy. It's your job to automate every other job out of existence. You're going to be just fine!"

He was happy with that answer, and if you can find a gig in automation, you'd probably be pretty happy with your prospects, too.

By the way, I have followed my own advice about gaining professional exposure to automation. In fact, my company, Prestige Economics, performs extensive research and data analysis for MHI, the material handling industry and trade organization in the United States. This is a multi-billion dollar industry that provides the physical equipment and technology that moves goods through the U.S. supply chain. These are the unsung heroes meeting the rising e-commerce needs of the U.S. economy.

Material handling also happens to be an industry that, in part, is focused on automation, robotics, last mile solutions, and transportation optimization. I have often joked that if my friend doing automated vehicle work will be the last guy on earth with a job, I'm just trying to be the second-to-last guy on earth with a job. The truth is, however, a bit more nuanced than that, as there are other critical industries that will also be difficult to automate — like healthcare.

Healthcare Jobs

Healthcare is likely to remain evergreen in a world of automation, since it is a more difficult sector to automate, and it involves high levels of human contact. In *Jobs for Robots*, I discussed the current state of (and near-term outlook for) the U.S. labor market. And healthcare came up a big winner for near-term job growth, job numbers, and top-end income.

If you are worried about automation, any industry where you need person-to-person contact is likely to remain relatively safe. Generally speaking, health care professions are likely to pay better than other high-contact service sector jobs like hairdressers, estheticians, and massage therapists.

Careers in health care are also likely to be solid for a long time to come, since the demographics of an aging U.S. population will necessitate expanding the ranks of front-line healthcare professionals, including personal care aides, registered nurses, and home health aides. In *Jobs for Robots*, I presented information about the positive outlook for healthcare job growth, and in the table below, you can see the current expectations for job growth by occupation in the U.S. economy. Healthcare is the clear winner.

Greatest Number of New Jobs by Occupation[2]

Most New Jobs

OCCUPATION	NUMBER OF NEW JOBS (PROJECTED), 2014-24	2015 MEDIAN ANNUAL PAY
Personal Care Aides	458,100	$20,980
Registered Nurses	439,300	$67,490
Home Health Aides	348,400	$21,920
Food Preparation/Service Workers	343,500	$18,910
Retail Salespersons	314,200	$21,780
Nursing Assistants	262,000	$25,710
Customer Service Reps	252,900	$31,720
Cooks	158,900	$23,100
General/Operations Managers	151,100	$97,730
Construction Laborers	147,400	$31,910
Accountants/Auditors	142,400	$67,190
Medical Assistants	138,900	$30,590
Janitors/Cleaners	136,300	$23,440
Software Developers	135,300	$98,260
Laborers	125,100	$25,010
Administrative Support Workers	121,200	$52,630
Computer Systems Analysts	118,600	$85,800
Licensed Practical/Vocational Nurses	117,300	$43,170
Housekeepers	111,700	$20,740
Medical Secretaries	108,200	$33,040

Source: BLS, Prestigo Economics LLC

FI | THE FUTURIST INSTITUTE

Project Management

As more tasks become automated, project and process management will become increasingly important. With an increasing number of potential uses of automation, the effective prioritization, optimization, and implementation of tasks will be required in order to ensure the highest dollarized value of automation solutions.

In the same way that word processors have necessitated that all professionals develop typing and secretarial skills, professionals of the future will need to have the project management skills of an MBA. Project management skills are even likely to exceed numeracy in importance. After all, even if robots can perform tasks and processes can be automated, it will be a very long time before they will be able to do this without guidance.

And that is where people come into the picture. If robots are doing all of the work, people are going to need to gain project management skills. Because people will need to manage the robots. There are three types of project management that people are already doing, but will need to do more of in the future:

Managing People — telling people which processes to implement in what order.

Managing Robots — telling robots which processes to implement in what order.

Managing People Who Manage Robots — telling the people what the top priorities are, so that they can tell the robots how to prioritize their workflow.

Without an ethical guidance system, and without an understanding of subjective corporate prioritizations, people will be a critical part of the puzzle for a long time to come. Robots can get things done — but only if instructed properly.

Surely, you have been in meetings and thought; why can't this just get done? It's like asking for directions, only to be told that you can't get there from here. To answer questions, to get things done, people need to be involved to help identify what the real question is, how to formulate that question, and how to make sure the question is not misspecified. Essentially, we all become MBAs. We all become management consultants. Execution goes to the robots. Planning, prioritizing, allocating resources, and directing activities remain with the humans.

Microentrepreneurs

Maybe IT, healthcare, and management are not for you. What then? Well, as the line from the film *The Untouchables* goes, "if you do not like the apples in the barrel, go pick one off the tree."

Disruption creates opportunity, and there are more opportunities than ever to start your own business. This is especially true at a time when the offices of today are at risk of becoming the museums of tomorrow. There has been an emergence of the so-called "gig economy." Arun Sundararajan, a professor at New York University, states in his book, *The Sharing Economy*, that there is the potential for a rise of "microentrepreneur[s]...self-employed workers who are empowered to work whenever they want from any location and at whatever level of intensity needed to achieve their desired standard of living."[3]

These new opportunities to be a microentrepreneur are supported by changes that also present greater chances for business owners to connect with the global capital markets than at any other time in history. Although Sundararajan writes that "crowd-based capitalism is still in its infancy,"[4] increased capital market access from crowdfunding presents significant opportunities. People can now turn hobbies into businesses by accessing the world via e-commerce, in-hand retail, and global capital markets.

There are challenges for this new and relatively untested initiative. And the future of crowdfunding and microentrepreneurs will depend heavily on investment profitability and liquidity. Investors will need to see that their privately-held crowdfunding investments can generate a return, and that these investments can be easily sold in a secondary marketplace. If those hurdles can be jumped, then there will be a lot more opportunities for "the little guy" to access global capital markets to build a business in the future.

It is now easier to reach an international market and access capital than ever before. If you ever dreamed of being a business owner, now might a good time to go for it! Before you go down that path, however, you might want to check out my book *Recession-Proof* for some of the most important secrets about the process and math behind starting and building a business.

Strategy #2: Learn Valuable Skills

As I discussed in *Jobs for Robots,* access to educational opportunities are much greater than at any other time in history. There are also a number of things you can do to enhance and add to your professional skill set to robot-proof yourself.

If you don't have time for a degree, get a certification. There's been huge growth in the number of professional designations available. I have a bunch of letters after my name, and other serious professionals do, too. These build skills, but they are also signaling devices. They signal to a future boss or client that you have knowledge, as well as drive, follow-through, and hunger.

Is there a particular computer program that you'd like to add to your resume? Download the free trial. The 30 days you'll get with that product is more than enough to learn the basics. Then add it to your resume. To include that skill on your resume, you don't need to be the world's expert on that product. Unless that computer program is a critical component of a job, you probably just need to have a basic competence and familiarity with the program, since most companies will train you once you are hired.

Audit a class at your local university. That means showing up to the lectures without being enrolled. You don't get credit or a grade, but you do gain the knowledge, and you can list it on your resume. At many universities, it's free. At others, there's a modest fee. The University of Texas at Austin, charges just $20 to audit an entire class—and that's at one of the country's top public universities. If you want to build your business skills, accounting or finance classes would be good options.

Strategy #3: Keep Moving

Once you've invested time in building the skills you need, make sure that your education and skills are matched to solid professional opportunities. It might require you to switch fields, relocate geographically, or just switch companies. Fortunately, you have some of the greatest tools in history available, to more easily make those transitions.

In-Hand Labor Market

There have been major changes in the way people look for work in recent years. At the dawn of the industrial revolution, if you lost your job, the only options were to ask the other people in your village for a job. Then, during the industrial revolution, you could have consulted a newspaper for listings. That remained relatively unchanged as the main source of job leads from the mid-1800s until around 2000, when job postings came online. Today, there are about 27 million new U.S. job postings per year, with 5 million online job postings online at any time.[5] Plus, companies like Indeed.com boast 20 million postings worldwide.

Can you imagine being stuck with whatever job openings happened to be in your hamlet, as opposed to being able to access 20 million job listings all over the world? And the number of job postings online, as well as our access to them, is only likely to increase over time. You now have a full in-hand labor market.

In the twelve months through December 2016, about 38 percent of workers experienced turnover.[6] Automation could accelerate the frequency of that turnover. But the good news is, if you aren't happy, at least you are unlikely to be stuck in a job forever.

We are Ready

One of the most important things about working is that it creates a purpose for people. People derive their identities from their profession, even though careers and professions change over time. Although jobs will be constantly changing in coming decades, people are far less tied to their occupations now as an identity than they were when the industrial revolution started.

Interestingly, the main three strategies I have proposed here are essentially the same strategies I would have given smiths and millers at the dawn of the industrial age:

Work in an Evergreen Industry

Learn Valuable Skills

Keep Moving

This is the benefit of historical experience. We know change is coming, and we are better prepared than ever before.

Put Tech on Your Side

Downside risks from further technological development can be mitigated, in part, by leveraging to the hilt all of the instantaneous and in-hand opportunities that we have: in-hand classroom, in-hand office, in-hand labor market, and even in-hand retail. Most people have this kind of access in the United States, and the Internet of Things will bring more people at home and abroad into the fold. This increased access should continue to create opportunities for individuals, like you.

You are Richer Than You Think

When people worry about the impact of robots and automation, it is important to keep in mind how large the wealth effect from technology has been. The greatest medieval king could not have imagined the amount of food available at any U.S. supermarket — let alone that the store might have an automated checkout. And the knowledge available at any time on your smartphone exceeds the content of everything lost in the Library of Alexandria.

There has been a growing debate about the usefulness of standard measures of the economy. In my opinion, Gross Domestic Product, or GDP, is very much a good measure of how the economy is growing. But it is not a good measure of the wealth that technology creates.

While I was working on this book, I had a lengthy discussion about the wealth effect of technology with Louis Borders, a founder of Borders Books, Mercury Startups, and HDS Global.[7] Louis pointed out to me that although wages seem to have remained stagnant, people's quality of life has gone up, because of the technology they have. After all, the average handheld device has more technology in it than the computers that launched the Apollo missions. So, if every person can own the equivalent of (what had been) a multibillion dollar computer for only a couple hundred dollars, that really improves their quality of life. Essentially, people are wealthier in a significant and meaningful way today — even if it doesn't show up in the data. You are much richer than any of your ancestors, thanks to the technology we have. The automation age could bring you even more wealth — but only if you are prepared.

END NOTES

1. *Artificial Intelligence, Automation, and the Economy. Executive Office of the President* (December 20, 2016). P 16. Retrieved February 11, 2017: https://www.whitehouse.gov/sites/whitehouse.gov/files/images/EMBARGOED%20AI%20Economy%20Report.pdf
2. U.S. Bureau of Labor Statistics. "Most New Jobs." *Occupational Outlook Handbook.* Retrieved February 11, 2017: https://www.bls.gov/ooh/most-new-jobs.htm
3. Sundararajan, A. (2016). *The Sharing Economy: The End of Employment and the Rise of Crowd-Based Capitalism.* Cambridge, Massachusetts: The MIT Press, p. 177.
4. Ibid., p. 202.
5. U.S. Bureau of Labor Statistics. "Job Openings and Labor Turnover — December 2016: Retrieved February 12, 2017: https://www.bls.gov/news.release/pdf/jolts.pdf
6. Conference Board. *Help-Wanted OnLine Data Series.* Retrieved February 17, 2017: https://www.conference-board.org/data/request_form.cfm
7. Thank you, Louis Borders, for allowing me to interview you for this book.

ABOUT THE AUTHOR

Jason Schenker is the Chairman of The Futurist Institute and the President of Prestige Economics. He is the world's top-ranked financial market futurist, and Bloomberg News has ranked Mr. Schenker one of the most accurate forecasters in the world in 38 different categories since 2011, including #1 in the world in 23 categories for his forecasts of the Euro, the Pound, the Swiss Franc, crude oil prices, natural gas prices, gold prices, industrial metals prices, agricultural commodity prices, and U.S. non-farm payrolls.

Mr. Schenker has written four books that have been #1 Best Sellers on Amazon: *Commodity Prices 101*, *Recession-Proof*, *Electing Recession*, and *Jobs for Robots*. Mr. Schenker is also a columnist for *Bloomberg View* and *Bloomberg Prophets*. Mr. Schenker has appeared as a guest and guest host on Bloomberg Television, as well as a guest on CNBC. He is frequently quoted in the press, including *The Wall Street Journal*, *The New York Times*, and *The Financial Times*.

Prior to founding Prestige Economics, Mr. Schenker worked for McKinsey & Company as a Risk Specialist, where he directed trading and risk initiatives on six continents. Before joining McKinsey, Mr. Schenker worked for Wachovia as an Economist.

Mr. Schenker holds a Master's in Applied Economics from UNC Greensboro, a Master's in Negotiation from CSU Dominguez Hills, a Master's in German from UNC Chapel Hill, and a Bachelor's with distinction in History and German from The University of Virginia. He also holds a certificate in FinTech from MIT, an executive certificate in Supply Chain Management from MIT, a graduate certificate in Professional Development from UNC, and an executive certificate in Negotiation from Harvard Law School. He is currently pursuing a certificate in Cybersecurity with NACD and Carnegie Mellon University. Mr. Schenker holds the professional designations CMT® (Chartered Market Technician), CVA® (Certified Valuation Analyst), ERP® (Energy Risk Professional), and CFP® (Certified Financial Planner).

Mr. Schenker is an instructor for LinkedIn Learning. His course on Financial Risk Management was released in October 2017. Additional courses will be forthcoming in 2018 and 2019.

Mr. Schenker is a member of the Texas Business Leadership Council, the only CEO-based public policy research organization in Texas, with a limited membership of 125 CEOs and Presidents. He is also a member of the 2018 Director class of the Texas Lyceum, a non-partisan, non-profit that fosters business and policy dialogue on important U.S. and Texas issues.

Mr. Schenker is an active executive in FinTech, as the founder of the foreign exchange FinTech startup Hedgefly, and as a member of the Central Texas Angel Network. Previously, he was the CFO of a private equity crowdfunding startup.

Mr. Schenker is the Executive Director of the Texas Blockchain Association, and he is also a member of the National Association of Corporate Directors, as well as an NACD Board Governance Fellow.

In October 2016, Mr. Schenker founded The Futurist Institute to help analysts and economists become futurists through a training and certification program. He is a Certified Futurist.

Visit The Futurist Institute:

www.futuristinstitute.org

FI THE FUTURIST INSTITUTE

ABOUT THE PUBLISHER

Prestige Professional Publishing, LLC was founded in 2011 to produce readable, insightful, and timely professional reference books. We are registered with the Library of Congress, and we are based in Austin, Texas.

Published Titles

Be The Shredder, Not The Shred

Commodity Prices 101

Electing Recession

Jobs For Robots

Robot-Proof Yourself

Future Titles

The Robot and Automation Almanac - 2018

Spikes: Growth Hacking Leadership

Commodity Prices 101: Second Edition

Recession-Proof: The Futurist Edition

The Brain Business

The Valuation Onion

— THE ROBOT AND AUTOMATION ALMANAC —

The Robot and Automation Almanac: 2018 is a forthcoming collection of essays by robot and automation experts, executives, and investors on the big trends to watch for in 2018. *The Robot and Automation Almanac: 2018* is being compiled by The Futurist Institute. It will be published by Prestige Professional Publishing in January 2018.

THE FUTURIST INSTITUTE

FI **THE FUTURIST INSTITUTE**

The Futurist Institute was founded in 2016 to help analysts become futurists by providing the content and context to take longer-term views on business opportunities, risk management, markets, and the economy. The Futurist Institute confers the Futurist and Long-Term Analyst™ (FLTA) designation and helps analysts become Certified Futurists™.

Current Courses

The Future of Work
The Future of Data
The Future of Finance
The Future of Transportation
Futurist Fundamentals

Future Courses

The Future of Energy
The Future of Leadership

Visit The Futurist Institute:

www.futuristinstitute.org

Prestige Professional Publishing, LLC

Austin, Texas

www.prestigeprofessionalpublishing.com

PRESTIGE
PROFESSIONAL PUBLISHING

ISBN: 978-1-946197-04-7 *Paperback*
    978-1-946197-05-4 *Ebook*

www.ingramcontent.com/pod-product-compliance
Lightning Source LLC
Chambersburg PA
CBHW070949210326
41520CB00021B/7120